BOOKWORMS

Properties of Matter

Mass and Weight

Arthur Best

Cavendish Square
New York

Published in 2019 by Cavendish Square Publishing, LLC
243 5th Avenue, Suite 136, New York, NY 10016

First Edition

Website: cavendishsq.com

Library of Congress Cataloging-in-Publication Data

Names: Best, B. J., 1976- author.
Title: Mass and weight / Arthur Best.
Description: First edition. | New York : Cavendish Square, [2018] |
Series: Properties of matter | Audience: K to grade 3. | Includes index.
Identifiers: LCCN 2018013852 (print) | LCCN 2018016091 (ebook) |
ISBN 9781502642271 (ebook) | ISBN 9781502642264 (library bound) |
ISBN 9781502642240 (pbk.) | ISBN 9781502642257 (6 pack)
Subjects: LCSH: Mass (Physics)--Juvenile literature. |
Weight (Physics)--Juvenile literature. | Matter--Properties--Juvenile literature.
Classification: LCC QC106 (ebook) | LCC QC106 .B47 2018 (print) | DDC 531/.14--dc23
LC record available at https://lccn.loc.gov/2018013852

Editorial Director: David McNamara
Copy Editor: Nathan Heidelberger
Associate Art Director: Alan Sliwinski
Designer: MeganAmetté
Production Coordinator: Karol Szymczuk
Photo Research: J8 Media

The photographs in this book are used by permission and through the courtesy of: Cover Vladimir Sukhachev/Shutterstock.com; p. 5 Thiago Santos/Alamy Stock Photo; p. 7 George Rudy/Shutterstock.com; p. 9 Ilya Akinshin/Shutterstock.com; p. 11 Food Collection/Getty Images; p. 13 Pirvu Marius/Alamy Stock Photo; p. 15 Charles M. Duke Jr./NASA, p. 17 Monty Rakusen/Cultura/Getty Images; p. 19 Krasowit/Shutterstock.com; p. 21 JSC/NASA.

Printed in the United States of America

Contents

Things are made of **matter**.

All things have weight.

You find weight with a **scale**.

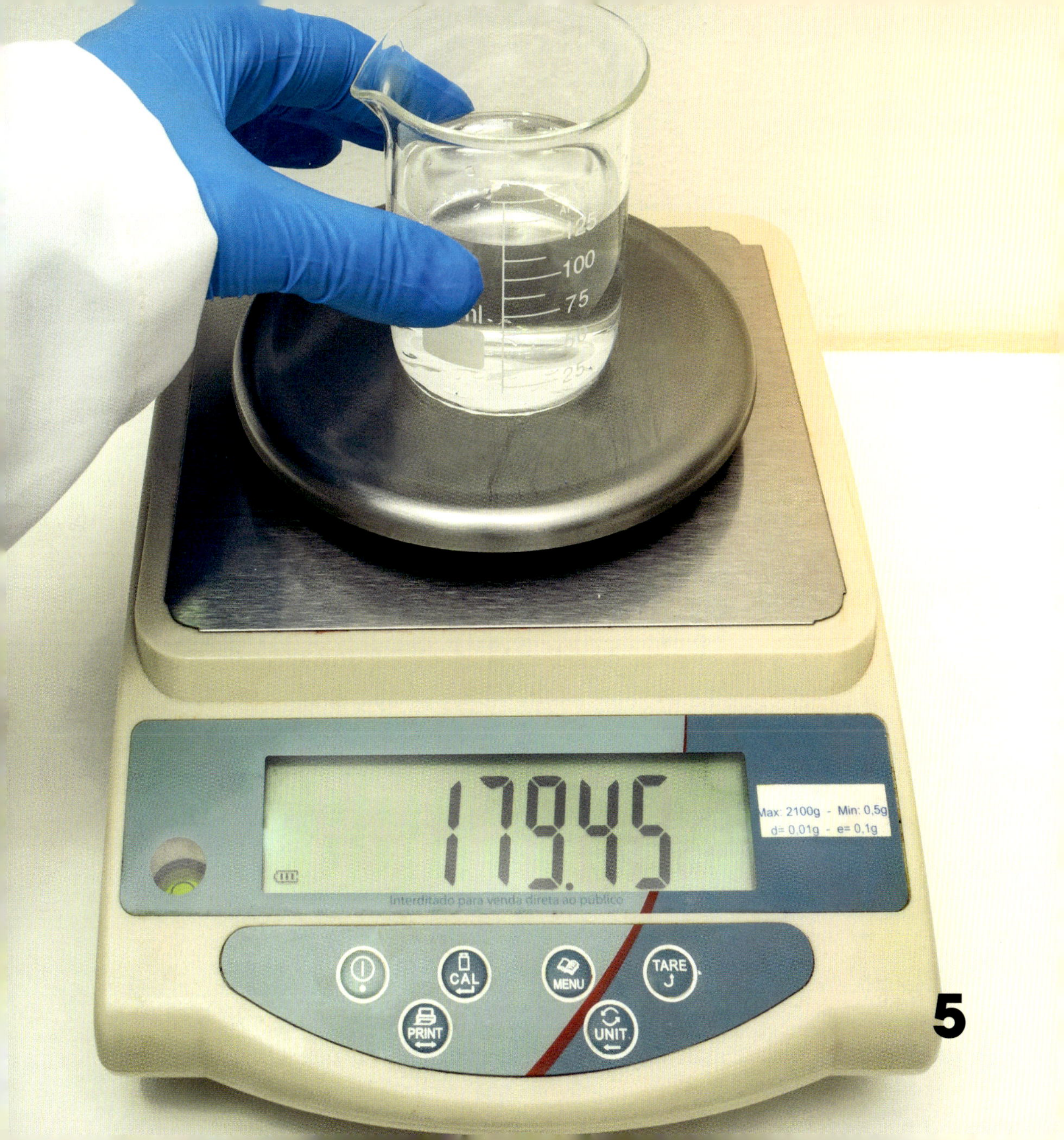
125
100
75
25
179.45
Max: 2100g - Min: 0,5g
d= 0,01g - e= 0,1g
Interditado para venda direta ao público
CAL
MENU
TARE
PRINT
UNIT

Some things are heavy.

They weigh a lot.

A bag of flour is heavy.

A bowling ball is heavy.

Some things are light.

They do not weigh a lot.

A feather is light.

A balloon is light.

Smaller things can weigh more than bigger things.

A bag of candy is heavy.

Cotton candy is light.

Earth has **gravity**.

Gravity pulls things down.

It gives things weight.

The moon has less gravity than Earth.

Things weigh less there!

Things are made of **atoms**.

Atoms are the smallest bits of matter.

You need a special **microscope** to see them.

Atoms are very small!

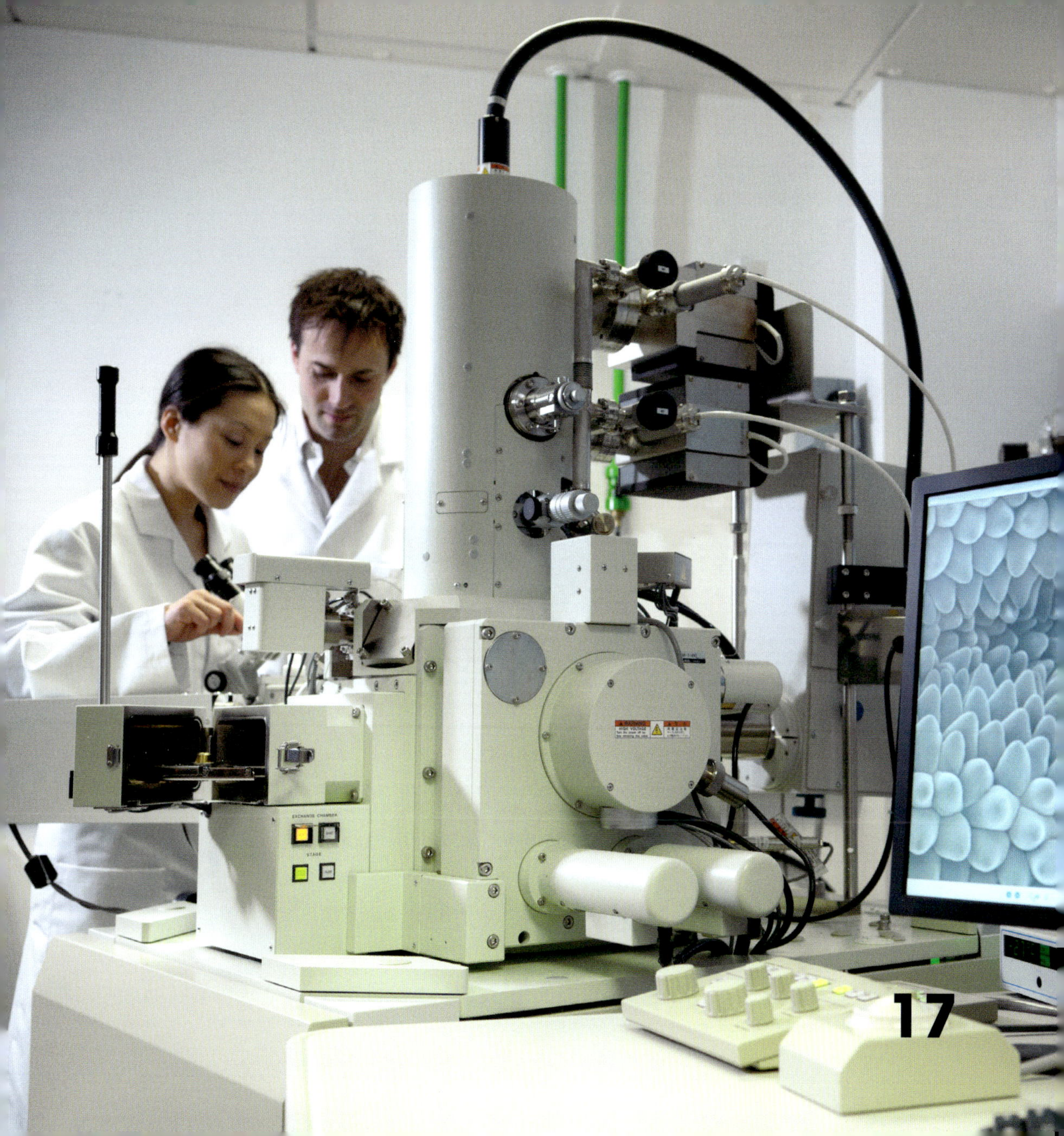

Things have a **mass**.

Mass is how much matter is in something.

Here is a bar of gold.

Its mass is made of gold atoms.

260373
10 oz
FINE
GOLD
999.9

Mass does not change.

Weight can change.

Outer space has less gravity.

You would weigh less there!

New Words

atoms (AT-ums) The smallest parts of matter.

gravity (GRAV-ih-tee) The force that pulls things down to the ground.

mass (MASS) How much matter is in something.

matter (MAT-er) What objects are made of.

microscope (MY-kruh-skope) A tool to see very small things.

scale (SCALE) A tool to find weight.

Index

About the Author

Arthur Best lives in Wisconsin with his wife and son. He has written many other books for children. He weighs 170 pounds.

About BOOKWORMS

Bookworms help independent readers gain reading confidence through high-frequency words, simple sentences, and strong picture/text support. Each book explores a concept that helps children relate what they read to the world they live in.